Mastering Cascade Control

George Buckbee, P.E.

PIDTutor offers many resources to help you to develop your process control skills. Visit the following sites on-line:

Main Web Site: http://www.pidtutor.com

Blog: http://wordpress.com/processcontrolguru

About the Author

George Buckbee, P.E. is the founder and president of PIDTutor. He has over 25 years' experience in the practical application of process control to a wide variety of process plants. George holds a B.S. and M.S. in Chemical Engineering, and is a registered professional engineer.

Table of Contents

Introduction to Cascade Control

Cascade control is a control strategy that uses two controllers working in concert. Figure 1 shows a typical cascade control system, where two instruments and a single valve are used to control the level in a tank. One controller manages the tank level by adjusting the inlet flow setpoint. The other controller handles fast disturbances in the inlet flow.

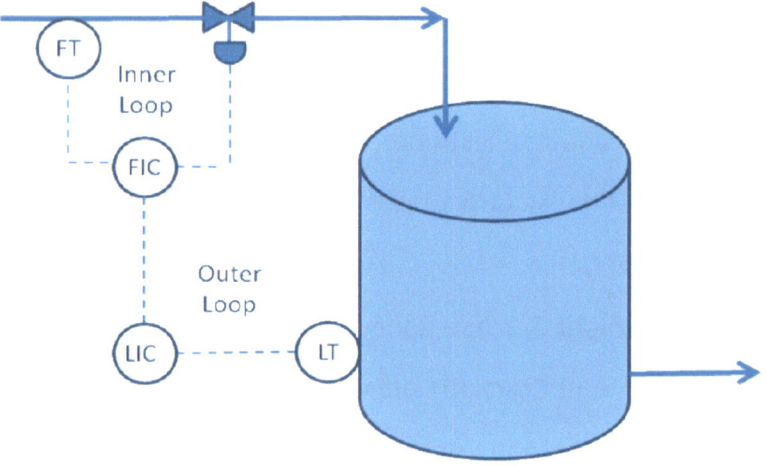

Figure 1. A Cascade Control System

The arrangement is called "Cascade" because the output from the level controller cascades to the setpoint to the flow controller. Figure 2 shows a control loop diagram for cascade control.

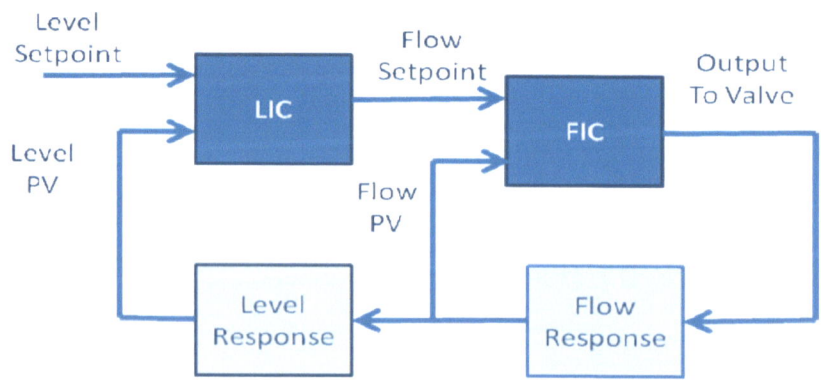

Figure 2. A Loop Diagram for a Cascade System

What You Will Learn:

In this book, you will learn about all aspects of cascade control, including:

- How cascade control works
- The costs and benefits of cascade control
- When to choose cascade as a control strategy
- When NOT to use a cascade strategy
- How to implement cascade control in a:
    - DCS
    - PLC
    - Set of Hardwired Controls
- Commissioning cascade systems
- Tuning methods for cascade controls
- How to troubleshoot cascade controls
- Operator training for cascade controls

How Cascade Control Works

The Basics

Some processes cannot be adequately controlled by a single PID control loop. For example, many temperature-control processes, like the one shown in Figure 3, respond very slowly. An upset in the steam supply pressure may result in a 20-minute excursion away from its temperature setpoint.

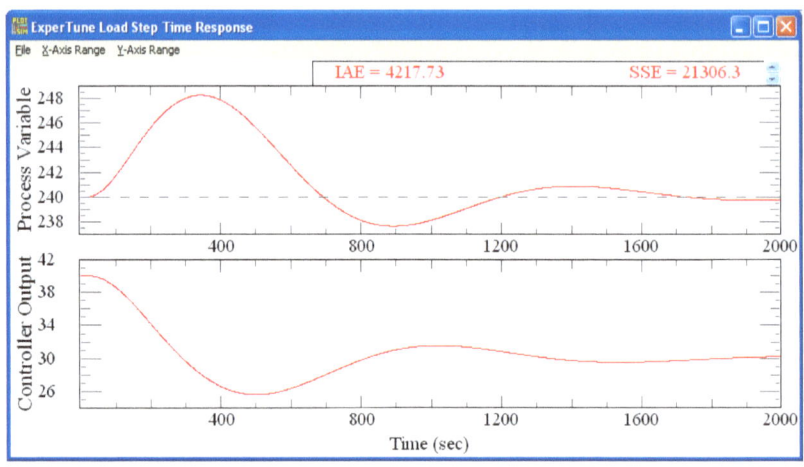

Figure 3. Temperature Upset without Cascade

By adding a steam flow sensor and flow controller, fast steam pressure upsets can be handled much more quickly. Figure 4 shows the resulting cascade arrangement.

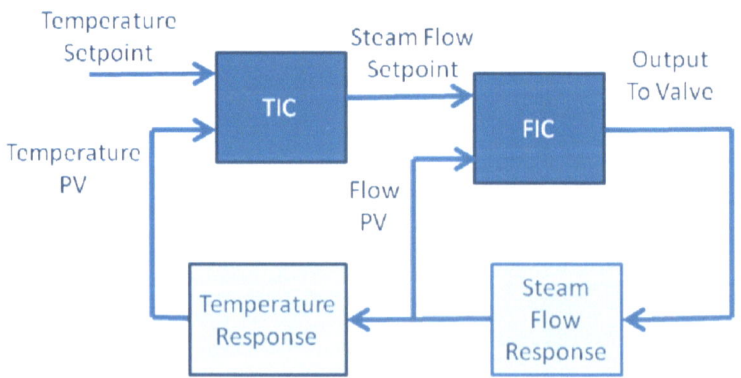

Figure 4. A Temperature Cascade Controller

The result is that steam upsets are managed very quickly, and temperature does not stray far from setpoint. Compare the results in Figure 5 from the original results in Figure 3. Notice that the time scale in Figure 5 is a short 100 seconds, compared with 2000 seconds in Figure 3.

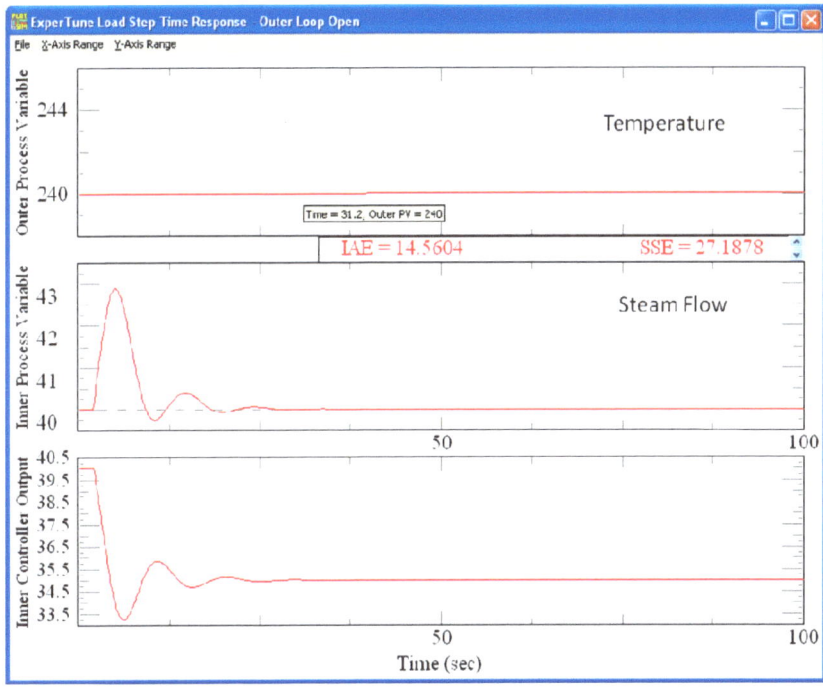

Figure 5. Temperature Upset with Cascade

Terminology

There are two loops in each cascade system. In this book, we refer to them as the "inner" loop and the "outer" loop. In the example above, the inner loop is the steam flow controller and the outer loop is the temperature controller.   Other terminology is sometimes used, as shown in Table 1. These terms are interchangeable. The Master Loop, Primary Loop, and Outer Loop all refer to the same loop.

Table 1. Cascade Loop Terminology

Outer Loop	Inner Loop
Primary Loop	Secondary Loop
Master Loop	Slave Loop

How it Works

The two control loops work together to manage the primary process variable (Temperature in the example).

The outer loop does not "know" that it is not connected to a valve. It is simply adjusting its output up and down to try to control the primary process variable. It has two modes of operation: MANUAL and AUTO.

The inner loop is controlling the secondary PV (ex. Steam Flow). It has 3 possible modes of operation: MANUAL, AUTO, and CASCADE. During CASCADE operation, the inner controller receives its setpoint from the outer controller.

Each loop, inner or outer, will most often use a P, PI, or PID control algorithm. However, it could even be another type of controller, like model-based controller or fuzzy logic. Regardless of the algorithm in use, all the same concepts apply.

The Costs and Benefits of Cascade Control

Costs

A Cascade controller system costs slightly more than a typical single-loop PID controller, primarily because it has two instruments.   A slight increase in engineering, design, installation, and commissioning effort will also be associated with Cascade controllers. The extent of the additional cost varies widely with the cost of the instrument. Table 2 compares the cost of a single, primary-only controller with the costs of a cascade system.

Table 2. Cascade Loop Costs

Single Control Loop	Cascade Loop
1 Instrument	2 Instruments
1 Valve	1 Valve
1 Controller	2 Controllers
In a Process Plant: $4,000 - $12,000	In a Process Plant: $6,000 - $18,000

So, all in all, a cascade system costs slightly less than double the cost of a single control loop. In most control loops, the majority of the costs are related to the valve and its installation. Adding the second loop costs less than the original control loop, because you do not need to buy another valve.

There are some smaller additional costs, such as increased operator training, which will be discussed later.

Benefits

The primary benefit of Cascade control is faster response to *some* process upsets. This, of course, leads to less variability, more stability, improved quality, and reduced operating costs.

A secondary benefit is that "some" control can be maintained even if the primary instrument fails, by putting the secondary loop in AUTO.

When to Choose Cascade as a Control Strategy

Certain process situations lend themselves directly to cascade control strategies. In other scenarios, however, cascade may provide little or no benefit. There are several criteria that define a good cascade application:

- An important process variable has slow dynamics, and is subject to upsets from faster-responding parts of the process.

- A secondary process variable provides a good measure of the disturbances, and can be independently controlled.

- Changes to the secondary process variable have a direct, if delayed, impact on the primary.

Typical Cascade Applications

Cascade control is appropriate for scenarios where the important primary process variable (PV) is slow-responding, and is subject to disturbances from other, faster-responding process streams. Some typical examples include:

- Heat Exchanger Temperature (outer loop) and Steam Flow Control (inner loop) as shown in Figure 4 and Figure 6.
- Tank Level (outer loop) and Feed or Discharge Flow (inner loop) as shown in Figure 7.
- Distillation Column Composition (outer loop) and Reflux Rate (inner loop) as shown in Figure 8.

In each case, look for the three criteria of a good cascade application:

1. A slow-responding outer loop provides a setpoint to a faster-responding inner controller

2. The secondary inner loop is more sensitive and faster-responding to upsets

3. The inner loop has an effect on the outer loop.

Also, notice that in each scenario, the outer loop is more important to the overall process outcome:

- Product temperature is more important than steam flow.

- Tank Level (residence time) is more important than the inlet flow rate.

- Distillation column composition is more important than the reflux rate.

These examples are explained in more detail below.

Example: Heat Exchanger and Steam Flow

Figure 6 shows a classic example of cascade control: a heat exchanger temperature cascaded to a steam flow controller.

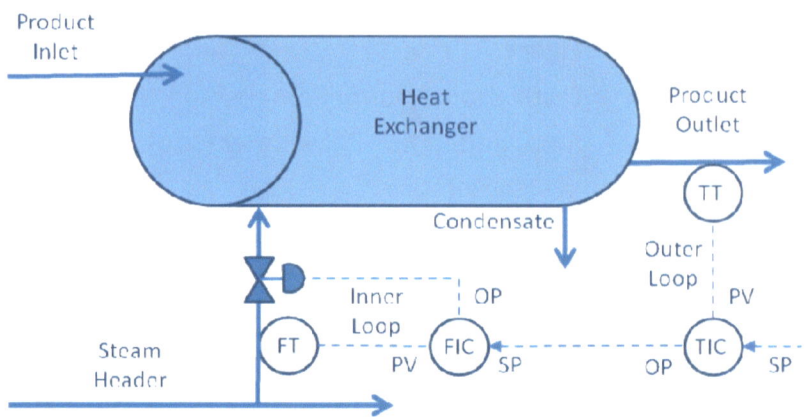

Figure 6. Example Temperature Cascaded to Steam Flow

The primary objective is to maintain a constant product temperature at the heat exchanger outlet. The temperature response is slow. Often, temperature loops may have 20 seconds of deadtime, and full response could take 10 minutes. This makes the loop difficult to control, and slow to respond to process upsets.

A major source of upsets is variation in the steam header. Many other processes, including other heat exchangers, may be drawing from the same header. Without the inner cascade, the temperature loop could see a lot of variation.

So we can add the steam flow control loop, a fast-responding loop, to the control strategy. We complete the cascade by providing the steam flow setpoint from the temperature controller's output.

This classic cascade application improves the speed of response to upsets, and ultimately reduces variations in the product outlet temperature.

Example: Tank Level Cascaded to Inlet Flow

Often, tank levels may be controlled with an inlet flow that comes from a common supply header. Supply header pressure may vary dramatically, because there are many different users. Adding a cascade loop, as shown in Figure 7, improves the response to these upsets, maintaining the desired flow rate.

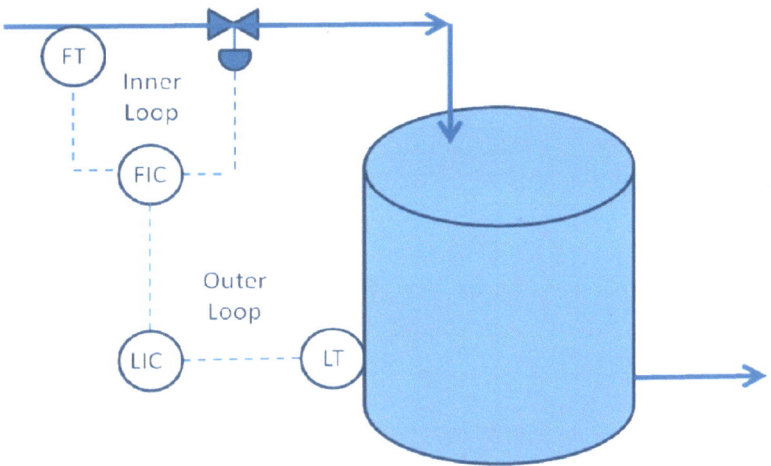

Figure 7. Tank Level Cascaded to Inlet Flow

The level controller provides a setpoint to the flow controller. A flow transmitter measures the inlet flow, and the fast-responding flow controller quickly adjusts the valve. This minimizes the impact of supply header pressure variations.

The level controller then makes adjustments to the flow setpoint to increase or decrease the inlet flow, as needed.

In other situations, tank level can be cascaded to the tank outlet flow. In that case, the level control action reverses, because the controller must INCREASE flow to DECREASE the tank level.

Example: Distillation Column Composition

Distillation columns have notoriously slow process dynamics. The final column composition may take 12 hours or more to fully respond to changes. Distillation columns may also have complex control strategies, with many different sets of cascaded loops. Figure 8 shows one such cascade example.

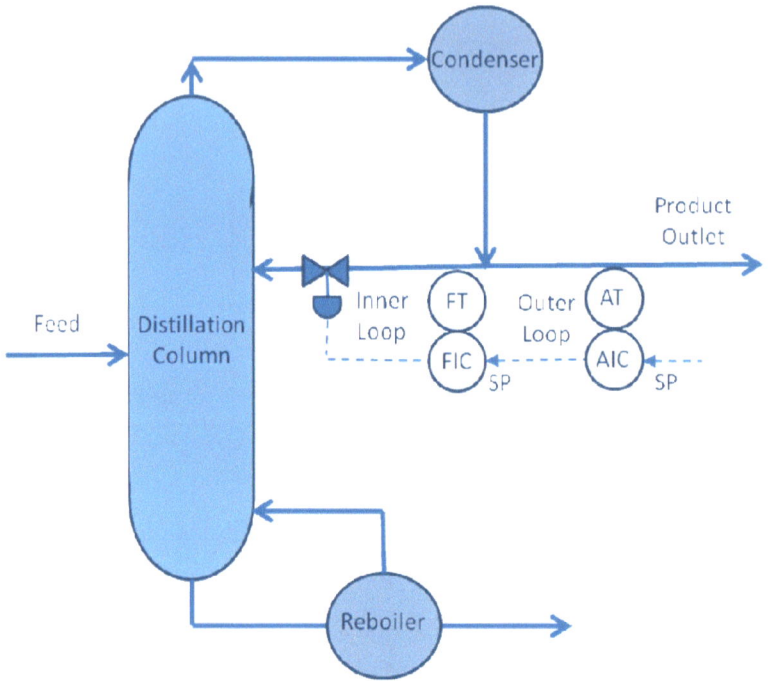

Figure 8. Distillation Column Composition Cascaded to Reflux Rate

This is, of course, a greatly simplified version of a column control strategy. Most columns have several other loops, often in cascaded or feed-forward arrangements.

In this case, the primary objective is to produce a product at a specified composition. The composition responds fairly slowly to changes in the column, for several reasons:

1. The measurement itself may be handled by a sampling device. Updated measurements may come only once per 5 minutes.

2. Columns have many internal trays. Each tray adds another dynamic time constant, or lag. The effect of multiple lags is to increase the apparent deadtime.

3. The internal reflux (liquid going downward, vapor coming upward) means that it may take hours for the column to stabilize after a single change or upset.

Again, we need to resort to a related fast-responding loop, to try to handle upsets more quickly. In this case, we have selected the reflux flow rate. This determines how much liquid is returned to the distillation column.

If the separation is not complete enough, more liquid is returned to the column. If we have over-distilled, less reflux is returned, and more product is drawn off.

Note that another cascade arrangement that is often employed is to cascade the condenser level to the product outlet flow rate.

When NOT to Choose Cascade

Cascade controllers are sometimes mis-applied to fast-responding control loops. When the primary process variable responds quickly, there is no benefit to adding a second cascaded loop. Keep it simple.

Examples of Cascade Mis-Application

Example 1: Positioners on Fast Loops. Control valve positioners are actually hardware control loops. A positioner is essentially a "valve position controller", designed to ensure that the valve is delivered to the value commanded by the loop controller.

When a positioner is applied to a fast loop, the first criteria for cascade is not met. ("An important process variable has slow dynamics…") If the outer loop is sufficiently fast by itself, there is simply no need for the inner loop. In this case, there is rarely a need for a positioner on a fast control loop, such as a liquid flow controller.

Example 2: Using Cascade for Ratio Control. It may be tempting to use the control output of one flow loop (the master) to establish the setpoint for another flow loop (the slave/follower) to maintain a flow ratio between them. See Figure 9 for a diagram of this strategy. When then master loop valve opens, the slave loop setpoint goes up proportionally, attempting to maintain a constant ratio between the two flows.

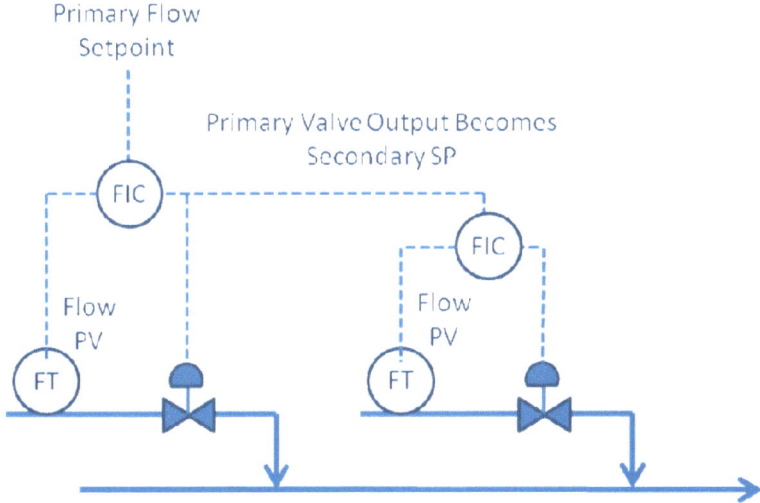

Figure 9. Cascade Used for Ratio Control (Not Recommended)

This strategy has several flaws:

1. The outer loop's valve is not perfectly linear. A 5% increase in control output is not likely to cause a 5% increase in flow.

2. The secondary (slave/follower) loop will always lag behind the outer loop. This difference in dynamics can be significant.

A better solution is to use the primary loop's PV to set the setpoint for the secondary loop. This eliminates the valve non-linearity issues. However, the dynamic lag problem will remain.

An even better solution is to use a ratio control strategy, where both flow rate setpoints are set by a common "Master Flow Rate".

How to Implement Cascade Control in Stand-Alone Controllers

In traditional stand-alone controllers, Cascade controls are configured electrically. Instead of wiring the primary control output to a valve, it is instead wired to "remote setpoint" terminals on the secondary (inner) controller, as shown in Figure 10.

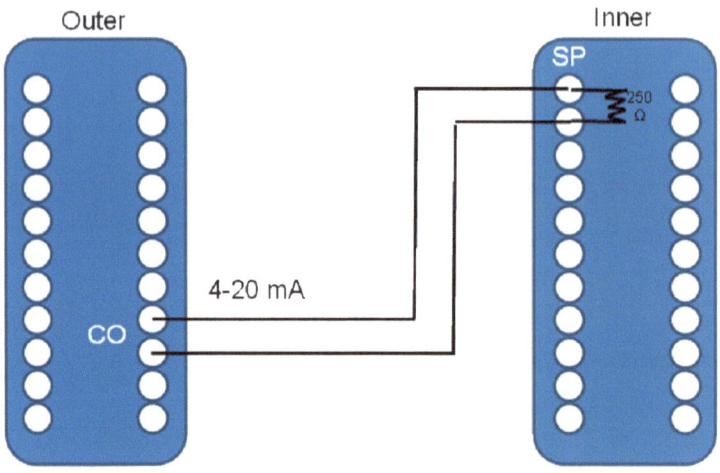

Figure 10. Wiring for Stand-Alone Controllers

This design is relatively simple to implement and maintain. The current output is typically a 4-20mA signal. The input, however, may require a 1-5 Volt signal. A 250-Ohm dropping resistor can be used to convert from the current signal to a voltage signal.

 NOTE: Operator training is required to ensure a smooth "bumpless transfer".

Bumpless Transfer

Bumpless Transfer refers to the procedure that is required when switching the inner controller from AUTO to CASCADE.

When the inner controller is switched to CASCADE, it immediately receives a new setpoint from the outer controller output signal. This results in an immediate "bump" to the inner loop setpoint. This is a "bumpy" transfer.

To ensure a "bumpless" transfer, follow this procedure:

1. Inner Loop in AUTO, Outer Loop in MANUAL.

2. Ensure the inner loop PV is at or near setpoint.

3. Set the Outer Loop control output signal to match the inner loop's setpoint. (in % of scale).

4. Switch the inner loop to CASCADE operation.

5. Switch the outer loop to AUTO.

The bumpless transfer procedure ensures that there is a smooth transition into cascade mode.

Also, note that stand-alone cascade controllers should be physically located adjacent to each other, to allow for the bumpless transfer procedure.

How to Implement Cascade Control in a Modern DCS

In a modern Distributed Control System (DCS), Cascade controllers can be configured in a straightforward manner. Figure 11 shows a typical implementation in a DCS.

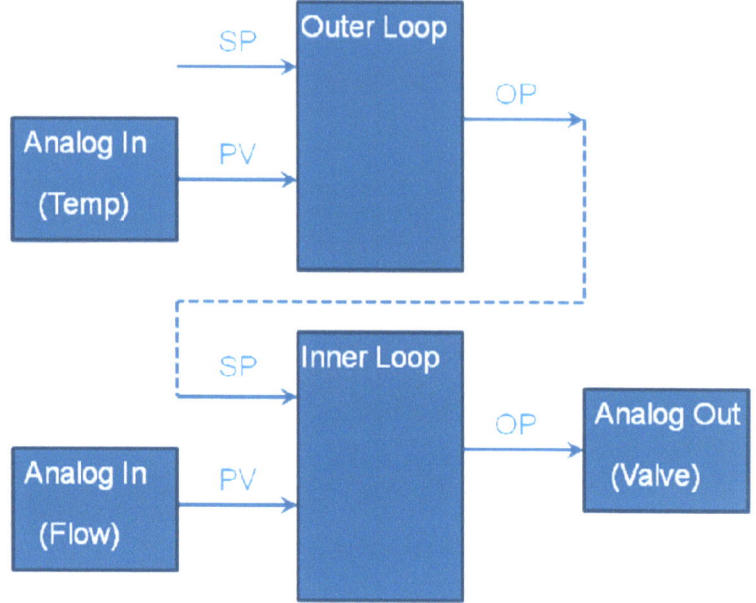

Figure 11. DCS Configuration of Cascade Control

The advantages of a DCS application are:

1. Mode transitions are typically handled automatically. For example, bumpless transfer can occur without any operator intervention.

2. Scaling can occur automatically (100% of outer loop output = 100% of inner loop setpoint scale).

Note that some DCS systems may place the loops in transitional modes, such as "INITIALIZE", during the transition from AUTO to CASCADE.

Of course, you should confirm the specifics of mode transition for your particular DCS.

How to Implement Cascade Control in a PLC

Implementation in a PLC is very similar to implementation in a DCS, although you often have to custom-develop the mode transition logic, or train operators to perform bumpless mode transitions. Figure 12 shows the logic for simple cascade implementation in a PLC.

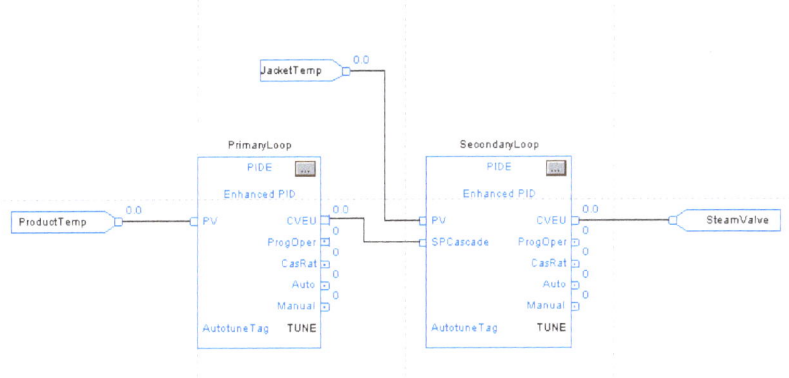

Figure 12. Simple PLC Configuration of Cascade Control

With PLC implementation, you must pay careful attention to several details, including:

1. Stick with the standard PID blocks. Do not "roll your own" PID block algorithm by writing custom ladder logic to perform the PID calculations. There are far too many pitfalls with writing a custom PID block application, including: integral windup, oversampling and loss of resolution, calculational instability.

2. Select the appropriate PID algorithm. Given a choice, you should use "D on PV".

3. For most PLC's - The actual scan rate for each PID block must match the scan rate configured in the block. Simply setting the value in the function clock does NOT force the block to scan at this rate.

Scheduling PID blocks in dedicated tasks is a preferred method.

4. Pay special attention to scaling. Some PLC's will work in engineering units, others in integer "counts". Make sure the outer loop's CO limits are equal to the inner loop's SP range.

5. Bumpless transfer and mode switching often require customized logic, and a customized user interface.

6. Choose the appropriate PID block options to allow/enable cascade. Some PLC's PID blocks have a separate switch to allow this. See the section on bumpless transfer.

7. Tune the controller in the proper units. Controller Gains most often are unitless, measured as (% output)/(% input). Some PID blocks have "relative gain" settings, that may multiply or divide controller gains by a factor of 10 or even 100.

How to Choose Cascade Controller Options

Inner Loop

The inner control loop is typically a fast-responding controller. Most often, it is a flow controller. The following options are recommended for these loops:

1. Derivative action is usually not required. If Derivative action is used, choose "Derivative on PV", not "Derivative on Error". The "Derivative on Error" selection will result in excessive valve movement as the D action responds to each setpoint change.

2. PV Tracking. When the inner loop is in MANUAL, this option forces the setpoint to follow the PV. This way, when the loop is returned to AUTO, there is a bumpless transfer.

3. Positioners are usually not required for these valves.

Outer Loop

For the outer, or master, control loop, consider the following options:

1. The outer loop is usually more important from a process perspective. So aggressive tuning is often desired. Adding derivative action may allow for more aggressive controller gain.

2. Output tracking, or "initialization". This is available on some DCS systems. When the outer loop is in manual, the initialization feature makes the control output of the outer loop match the setpoint of the inner loop. This feature is not available on most PLC systems.

Commissioning Cascade Systems

Procedure

One great advantage of cascade control is that it can be operated in part or in whole. The commissioning process takes advantage of this.

Commission the Inner Loop

Start by commissioning the inner loop:

1. Ensure all wiring/addressing is installed as per specification
2. Test/Stroke the Valve
3. Test the Instrument
4. Tune the inner control loop. (see Tuning Cascade Controllers, below.)

Once the inner loop is tuned, it can be operated in MANUAL or AUTO modes. However, it should not be operated in CASCADE until after the outer loop commissioning process is completed.

Commission the Outer Loop

For the outer loop, follow this commissioning process:

1. Ensure wiring/addressing is installed as per the design.

2. Test the instrument.

3. With the **outer** loop in MANUAL, attempt to perform a bumpless transfer, placing the **inner** loop into CASCADE. Note: Be prepared to switch the **inner** loop back to AUTO or MANUAL if something unexpected happens.

4. Tune the outer control loop. (see Tuning Cascade Controllers, below.)

Operator Training

Of course, operators must be trained in Cascade applications. The following key points are important for operator training:

- In the event of an emergency, instrument failure, or control problem, control of the **inner** loop is most expedient. It is the inner loop that has the control valve, and therefore it has direct impact on the process. Emergency procedure may call for placing the inner loop in MANUAL, and moving the valve to a safe operating position.
- The outer loop does not have a valve. Placing the outer loop's output to zero does NOT place the system in a safe mode.
- When the inner loop is in MANUAL or AUTO, the outer loop controller has no effect.
- Keep in mind that the outer loop output position provides a setpoint, not a valve position. The inner loop valve may be in <u>any</u> position, regardless of the outer loop's control output position.

- Depending on the control system, the operator may be required to perform a specific procedure to ensure bumpless transfer.

- In many user interfaces (HMI, DCS) the outer loop's output value may be displayed in different engineering units than the inner loop's setpoint, *even though they are the same signal*.

Tuning Cascade Controllers

Tuning of cascade controllers is poorly understood in most plants. For this reason, many cascade control systems tend to "fight" each other, as illustrated in Figure 13.

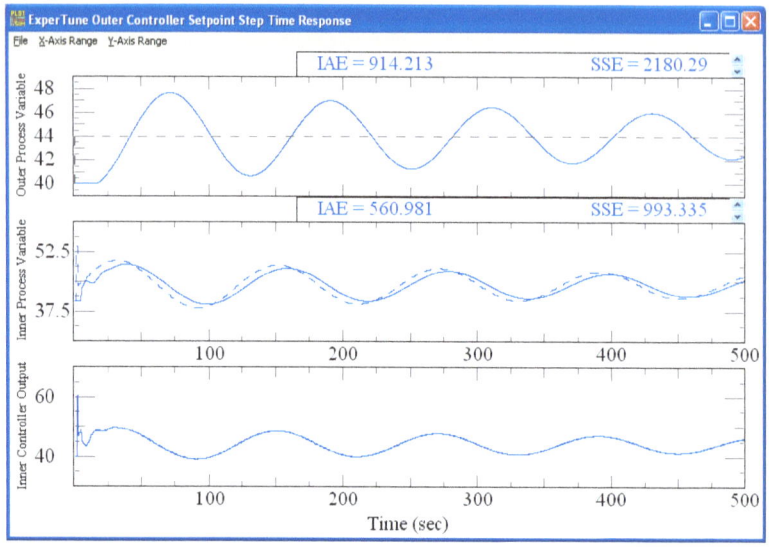

Figure 13. Poorly-Tuned Cascade Controllers (SP Response)

To ensure that the cascade controllers do not fight each other, the outer controller must be tuned more slowly than the inner controller. This allows for the inner controller to get its work done without interference from the outer controller.

The goal is to have two cascade loops that support each other. A properly tuned cascade system is shown in Figure 14.

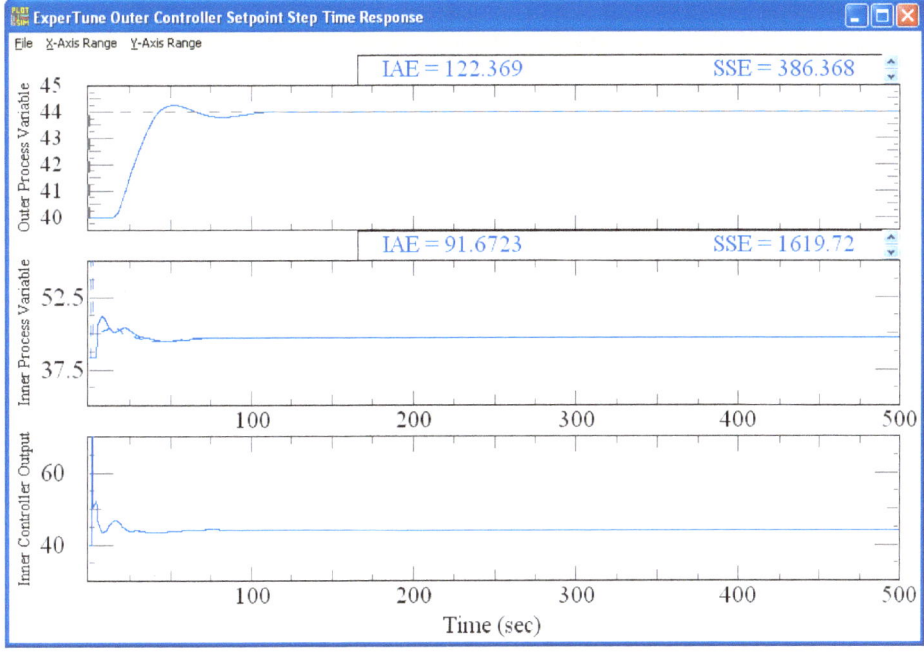

Figure 14. Correctly-Tuned Cascade Controllers

Cascade Tuning Procedure

A structured approach to tuning is recommended. Most structured approaches require bump testing to determine process models. When tuning Cascade systems, bump tests should include:

- Several bumps, throughout the range of operation for each loops.

- Steps in both the upward and downward directions.

The trick to tuning cascade loops is that the inner loop must be at least 3 times faster than the outer loop, to avoid unnecessary interactions and oscillations. The following procedure will ensure that you can avoid this common problem.

1. Place both loops in Manual.

2. Perform bump tests on the inner loop.

3. Tune the inner loop. Generally, you should choose aggressive tuning for this loop.

4. Return the inner loop to CASCADE mode. Leave the outer loop in MANUAL.

5. Perform bump tests on the outer loop. (Note that each bump is effectively a setpoint change to the inner loop.)

6. Tune the outer loop so that it's response time is at least 3 times larger than the response time of the inner loop.

How to Determine the Loop Response Time

Since response time is so critical to the cascade tuning procedure, we need a good way to measure it. There are several possible methods: closed-loop time constant, frequency-response method, or rise-time.

Rise-Time Method

The simplest, but least accurate, method is to measure each loop's rise time in response to a setpoint change, as shown in Figure 15. In a real-world control loop, noise makes it difficult to determine the actual rise time.

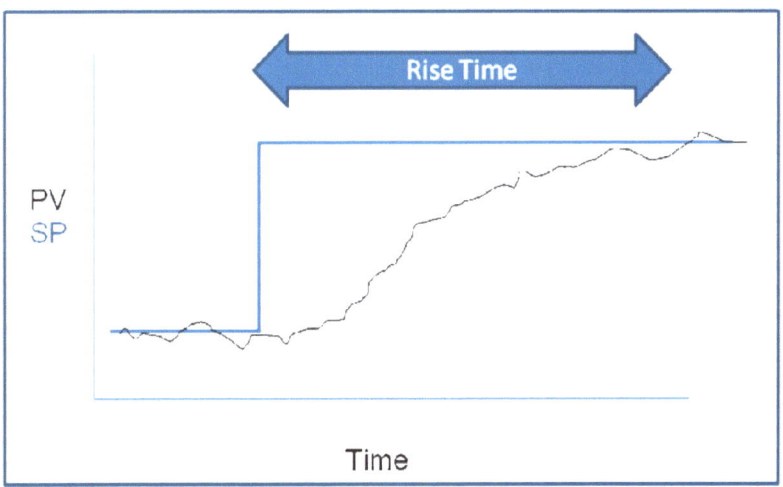

Figure 15. Determining Closed-Loop Rise Time

Closed Loop Time Constant Method

This is a measure of the time required for the loop to respond to a setpoint change. You can estimate the closed-loop time constant as the time required for the PV to reach 63% of the setpoint change, as shown in Figure 16.

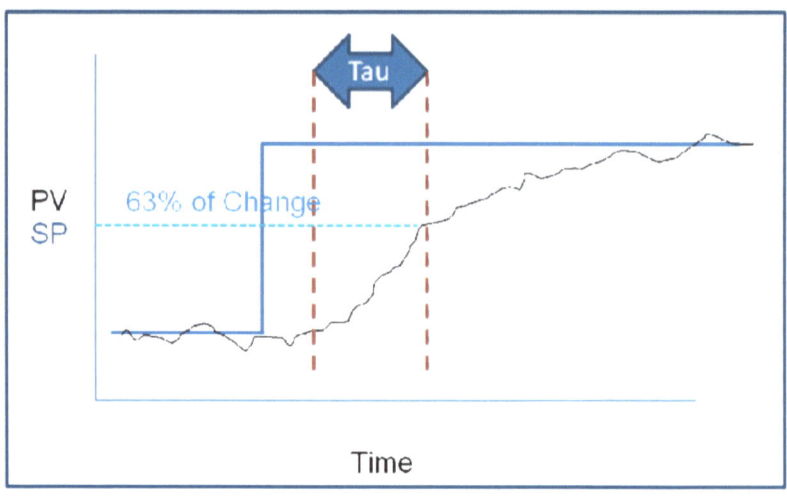

Figure 16. Determining Closed-Loop Time Constant

Frequency-Response Method

The most accurate method is the frequency-response method. Using this method requires software to analyze the frequency response of each loop. Use the plot of the loop's Amplitude Ratio plot to determine its corner frequency, as shown in Figure 17.

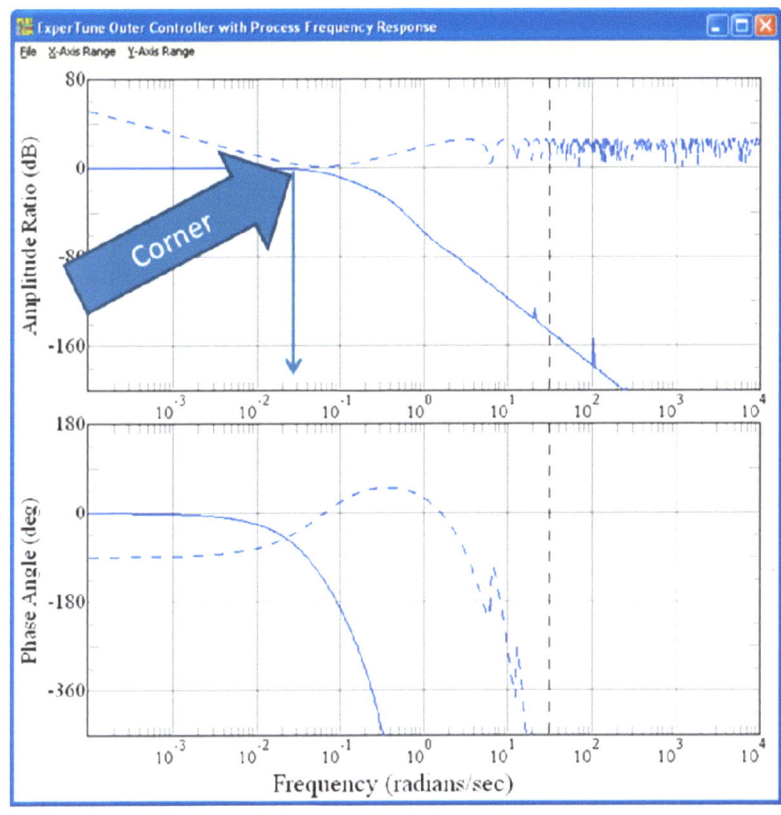

Figure 17. Determining Corner Frequency

The corner frequency is a good reference point to determine how fast the loop responds.

Adjusting the Response Time

The simplest way to adjust response time in a cascade system is to de-tune the outer loop. This can be accomplished by decreasing the controller gain (or increasing Proportional Band).

Generally, it is a safer approach to make the outer loop more sluggish, rather than to make the inner loop more aggressive.

Troubleshooting Cascade Controllers

A great strength of Cascade controllers is that they are made of two simple PID controllers. There are, however, some special problems that are unique to Cascade controllers. These problems, their causes and corrective actions are shown below in the "Trouble-Cause-Correction" table.

Trouble	Cause	Correction
Both loops cycle after any upset.	Interaction between controller tunings.	Tune both inner and loops, using the Cascade tuning procedure.
Process upset when inner loop is put in CASCADE	Loops are not configured for bumpless transfer.	Configure the loops for initialization and bumpless transfer.
Excessive fast movement of the inner loop's control valve.	Use of derivative action.	Configure inner loop for "Derivative on PV", or eliminate the use of Derivative action.
Outer loop response is too slow.	Inner loop tuning may be too slow.	Ensure inner loop is tuned for fast, or aggressive response.
No response to SP changes in the outer loop.	Inner loop is not in CASCADE, or not initialized.	Place inner loop in CASCADE. Check that initialization is completed on mode change.
	Outer loop in MANUAL	Place Outer Loop in AUTO

Conclusions

The Cascade control scheme provides fast disturbance rejection for an important primary process variable. Cascade offers the following advantages:

- It responds more quickly to process upsets.

- It uses two simple PID control loops.

- No special equipment of software required.

- It is easily maintained.

- It can be implemented in stand-alone controllers, PLCs, or DCS systems.

- Alternate operating modes allow safe operation in case of partial failure. The inner loop may be placed in AUTO or MANUAL.

Proper control performance of the Cascade system depends upon proper configuration, tuning, and operator training.

References & Further Reading

Buckbee, George, <u>Mastering Split-Range Control</u>, PIDTutor.com, Pennsylvania, 2009.

Liptak, Bela, <u>Instrument Engineer's Handbook: Process Control</u>, Chilton Book Company, Radnor Pennsylvania, 1995

Shinskey, F.G., <u>Process Control Systems</u>, 4th Edition, McGraw-Hill, 1996

Other Books by PIDTutor

Other process control books are available from PIDTutor. Visit PIDTutor.com for more details.

www.ingramcontent.com/pod-product-compliance
Lightning Source LLC
Chambersburg PA
CBHW041610180526
45159CB00002BC/793